The Seven Stars Of Orion

James B. Rogers

Table of contents

Introduction

For as long as there has been humanity, the Orion constellation, a celestial masterwork that lights up the night sky, has fascinated astronomers and storytellers alike. One of the most well-known constellations, Orion is located on the celestial equator and can be seen from almost every place where people live on Earth.

This celestial arrangement presents the unique profile of a great hunter, with three brilliant stars—Alnitak, Alnilam, and Mintaka—comprising the famed "Belt of Orion." Beneath the belt, Orion's feet are represented by the bright stars Saiph and Rigel, while his left shoulder is identified by the dazzling star Betelgeuse.

These bright stars define the form of the constellation, offering a mesmerising backdrop for the many cosmic stories that have been spun around it.

Orion's mythology is rich and diverse, spanning several civilizations. Greek mythology describes Orion as a big, expert hunter who is often connected to the gods. People all throughout the world may see the constellation because of its location in the night sky, which crosses the celestial equator and makes it visible from both the northern and southern hemispheres.

Orion is useful for celestial navigation in addition to its legendary charm. Orion's Belt is oriented to point to various celestial landmarks, which helps contemporary astronomers and ancient seafarers find their way around in the expanse of the night sky.

Astronomically speaking, this constellation gains additional importance from the Orion Nebula, which is situated close to the hunter's sword and serves as a stellar nursery where new stars are created. It is a stellar laboratory for researching star formation and the first phases of stellar development because of its closeness to Earth.

Through the pages of "The Seven Stars of Orion," the constellation takes on the role of both guide and storyteller, beckoning us to investigate the interwoven domains of mythology, science, and cultural symbolism that come together to form the enthralling night sky.

Chapter One

The History and Mythology of Orion

Greek mythology depicts Orion, the hunter, as the most attractive man alive. He was the child of King Minos of Crete's daughter Euryale and the sea deity Poseidon. Orion is portrayed in Homer's Odyssey as being very tall and carrying an indestructible bronze club.

In a tale, Orion developed feelings for the seven sisters, the Pleiades, who were the offspring of Pleione and Atlas. Zeus gathered them up and set them in the skies as he began to pursue them. Located in the Taurus constellation, the well-known Pleiades star cluster serves as a representation of the Pleiades. At night, Orion is still seen pursuing the sisters across the sky.

In a different tale, Orion harboured feelings for Merope, the beautiful daughter of King Oenopion, but she refused to reciprocate his feelings.

He had too much to drink one night and attempted to impose himself on her. Furious, the monarch closed Orion's eyes and exiled him from his homeland, the island of Chios.

Feeling pity for the sightless, nomadic Orion, Hephaestus donated one of his attendants to lead and serve as his eyes. Eventually, Orion came across an oracle that promised to restore his eyes if he headed east towards daybreak. When Orion followed through, his eyes were miraculously restored.

Originating in Sumerian mythology, the constellation Orion is associated with the story of Gilgamesh. The Sumerians connected it to their hero's battle against Taurus, the bull of heaven. Orion was referred to as URU AN-NA, which means "the light of heaven." GUD AN-NA, or "the bull of heaven," was their term for the Taurus constellation.

Although Orion is sometimes shown as being charged by a bull, there are no tales in Greek mythology that relate such a story. When defining the constellation, the Greek astronomer Ptolemy described the hero with a club and lion's pelt, both of which are commonly associated with Heracles, although there is no proof in mythology literature of a direct tie between the constellation and Heracles.

Nonetheless, there are at least indications of a potential relationship between the two as Heracles, the most well-known of the Greek heroes, is symbolised by the much less noticeable constellation Hercules, and since one of his missions was to capture the Cretan bull.

The majority of mythology about Orion's demise centre on a scorpion, however the details vary throughout mythologists. A legend has it that Orion told the goddess Artemis and her mother Leto he could slay any beast on the planet. After hearing him, the Earth Goddess dispatched a scorpion, which bit the giant.

In a different tale, Artemis was the one who sent the scorpion when he attempted to push her. In a different story, Orion suffered a sting while attempting to protect Leto from the scorpion. Every tale of Orion's demise has the same conclusion: Orion and the scorpion were positioned on opposing sides of the sky, allowing Orion to drop below the horizon in the west as a means of escaping the scorpion when the constellation Scorpius appears in the sky.

There is another tale, however, that does not include a scorpion. It tells of Artemis, the goddess of hunting, falling in love with the hunter and being challenged by her brother Apollo to use her bow and arrow to strike a little target in the distance in order to prevent her from breaking her chaste vows.

She nailed it on her first attempt, not realising it was Orion, who was having fun in the water. She set Orion among the stars, devastated by his death.

In many civilizations, Orion is a well-known constellation. The stars that make up Orion's Belt and Sword are commonly referred to as the Pot or the Saucepan in Australia.

The three stars in Orion's Belt are referred to as Drie Konings (the three kings) or Drie Susters (the three sisters) in South Africa. In Spain and Latin America, the stars are nicknamed Las Tres Marías, or The Three Marys.

Babylonians recognised Orion as MUL.SIPA.ZI.AN.NA or The celestial Shepherd (The True Shepherd of Anu) in the Late Bronze Age and identified the constellation with Anu, the deity of the celestial regions. Egyptians identified it with Osiris.

Orion was also associated with Unas, the last Pharaoh of the Fifth Dynasty, who was credited with becoming powerful and passing on his authority by allegedly consuming the flesh of his opponents and the gods themselves. Legend has it that Unas ascends across the sky to become Orion, or the star Sabu.

Some of the biggest pyramids, like the ones at Giza, were constructed to mimic the pattern of stars in the constellation Osiris because it was thought that pharaohs were turned into the god after death. The Great Pyramid's King's Chamber air shaft was oriented to align with Zeta Orionis, or Alnitak, the easternmost star in Orion's Belt, to facilitate the transformation.

The New Fire ceremony, which the Aztecs conducted to delay the end of the world, began when the stars of Orion's Belt and Sword rose in the sky. These stars were known to the Aztecs as the Fire Drill.

Orion is associated with Nimrod, a renowned hunter and the father of the twins Hunor and Magor, sometimes referred to as Hun ad Hungarian, in Hungarian mythology.

According to Scandinavian customs, the constellation is linked to the goddess Freya and is referred to as Frigg's Distaff, or Friggerock, after the spinning instrument that she employed. The constellation was named after the legendary Chinese hunter and warrior Shen.

From the second millennium BC comes another old tale. The Hittites (a Bronze Age people of Anatolia, the area including most of present-day Turkey) connected the constellation with Aqhat, a famed mythological hunter. He was admired by the battle goddess Anat, but she attempted to take his bow when he refused to give it to her. But the guy she dispatched to retrieve the bow failed miserably at the task, losing Aqhat and dumping the bow into the ocean. This explains why the constellation is said to disappear behind the horizon for two months in the spring.

Chapter Two

Rigel, Betelguese, Bellatrix, Saiph

In the northern hemisphere of the sky, Orion is a well-known constellation that is visible in the winter. It is one among the 88 constellations that exist today; the scientist Ptolemy counted 48 constellations in the second century. It has the name of a Greek mythological hunter.

The seven brightest stars in Orion create a recognisable asterism, or pattern, in the night sky that resembles an hourglass. The three stars of Orion's Belt—Alnitak, Alnilam, and Mintaka—are situated in the middle of a vast, approximately rectangular shape formed by the four stars Rigel, Betelgeuse, Bellatrix, and Saiph.

Rigel

The constellation's brightest star is Rigel. Additionally, it is the sixth brightest star in the sky with an apparent magnitude of 0.18. It is almost always brighter than Betelgeuse, Alpha Orionis, although not having the label alpha.

Actually, Rigel is a stellar system made up of three stars. Since 1831, or maybe even earlier, when F. G. Struve took the initial measurements, it has been recognised to be a visual binary. A shell of released gas envelops Rigel.

The Arabic expression Riǧl ǧawza al-Yusra, which translates to "the left foot of the central one," is where the word Rigel originates. Rigel denotes the left foot of Orion. "The foot of the great one," or iǧl al-ǧabbār, is another Arabic term for the constellation. This sentence is the source of the other two alternate names for the star, Algebar and Elgebar.

Blue supergiant Rigel is. It is located 772.51 light years away and has spectral class B8lab. It contains 17 solar masses and is 85,000 times brighter than the Sun.

It is categorised as a somewhat irregular variable star because during a period of 22 to 25 days, its brightness fluctuates between 0.03 and 0.3 magnitudes.

The principal component in the system, Rigel A, is 500 times brighter than Rigel B, which is also a spectroscopic double star. Rigel B has a magnitude of 6.7. It is made up of two main sequence stars of the B9V type that revolve around a shared centre of gravity once every 9.8 days.

Beta Orionis Rigel is linked to many neighbouring dust clouds that it illuminates. The most well-known is IC 2118, also referred to as the Witch Head Nebula. It is a weak reflection nebula in the constellation Eridanus that is situated around 2.5 degrees northwest of Rigel.

Member of the Taurus-Orion R1 Association is Rigel. It was regarded by some to be an outlying member of the Orion OB1 Association, a collection of several dozen hot giants belonging to the spectral categories O and B, found in the Orion Molecular Cloud Complex. The star is too near to us, however, to really belong to that specific stellar association.

Rigel's age is just a few million years. It will eventually mature into a crimson supergiant that resembles Betelgeuse greatly.

Betelgeuse

The eighth-brightest star in the sky and the second-brightest star in Orion is called Betelgeuse. It is a red supergiant that is a member of the M2Iab spectral class. Betelgeuse is categorised as an intermediate luminous supergiant, meaning it is not as brilliant as other stars like Deneb in the constellation Cygnus, according to the suffix -ab.

The star is probably no longer classified since some recent results indicate that it emits more light than 100,000 Suns, making it more brilliant than most stars in its class.

The star is around 643 light years away and has an apparent magnitude of 0.42. Among known stars with the highest luminosities is Betelgeuse. It is -6.05. in absolute magnitude.

Known as Alpha Orionis, Betelgeuse is among the biggest stars in the universe, with an apparent diameter ranging from 0.043 to 0.056 arc seconds.

The star seems to sometimes change form and has a vast envelope around it due to a significant mass loss, making an exact measurement challenging.

It is determined that Alpha Orionis is a semi-regular variable star. With an apparent magnitude that ranges from 0.2 to 1.2, Betelgeuse may sometimes eclipse Rigel, its brilliant neighbour.

On the other hand, this seldom ever occurs. Sir John Herschel originally mentioned the star's brightness variations in his 1836 book Outlines of Astronomy.

Though not particularly ancient for a red supergiant, Betelgeuse is estimated to be about 10 million years old. However, because to its massive mass, the star is assumed to have grown quite quickly.

Its existence will probably terminate in a supernova within the next million years. When it does, it will be visible in the sky both during the day and at night. The supernova would be the brightest supernova ever seen in history at its present distance from the solar system, with a brightness comparable to that of the Moon.

It's unclear where the name Betelgeuse originated. The last component, -elgeuse, is derived from al-Jauzā', the Arabic name for the constellation. This term was originally feminine in ancient Arabian traditions and means approximately "the middle one."

The most widely accepted explanation for the name Betelegeuse is that it is a corruption of the Arabic phrase Yad al-Jauzā', or the Hand of al-Jauzā', which is to say, the hand of Orion.

This corruption occurred during the Renaissance when the Arabic letter for y was mistaken for the Arabic letter for b, resulting in the name Bait al-Jauzā', or "the house of Orion." This ultimately resulted in the current name of the star, Betelgeuse.

Bellatrix

Slightly less brilliant than Castor in Gemini, Bellatrix, commonly referred to as the Amazon Star, is the third brightest star in Orion and the 27th brightest star overall. The Latin term for "the female warrior" is where its name originates. It is located around 240 light years away and has a typical apparent magnitude of 1.64.

Bellatrix is a blue-white massive star that is hot and bright and is categorised as an eruptive variable. Its magnitude fluctuates between 1.59 and 1.64. The star belongs to the spectral class B2 III.

One of the hottest stars that is visible to the unassisted eye is this one. It has eight or nine solar masses and emits light 6,400 times more than the Sun does.

Bellatrix will develop into an orange giant and then a gigantic white dwarf in a few million years.

Gamma Orionis was used as a standard for stellar brightness, one against which other stars were measured and evaluated for variability, before its own variability was established.

Saiph

Saiph, often referred to as Kappa Orionis, is a supergiant star in the constellation Orion, The Hunter, with a bright brightness. Part of the constellation, Saiph is a massive star. The star is known by its traditional/proper name, Saiph, and its Bayer designation is Kappa Orionis.

Saiph hue is blue because to the spectral class (Bo.5Iavar), indicating that the star is hotter than our star and among the hottest in the universe.

The temperature of Saiph is between 10,000 and 30,000 Kelvin. Since we don't know the precise temperature, we may infer from the spectrum type (Bo.5Iavar) that Saiph's surface temperature is between 10,000 and 25,000K based on the Harvard University comments. To put this in perspective, Google states that the Sun's temperature is around 5,778 Kelvin.

According to Hipparcos 2007 apparent magnitude, Saiph is the sixth brightest star in Orion and the 58th brightest star overall in the night sky.

Without the need for binoculars or a telescope, Saiph is a naked eye star that may be viewed on a clear night. The Saiph distance from Earth is 647.15 light years, or 221.24 parsecs, based on a parallax of 5.04.

A beta cephei variable star is Saiph. Stars that fluctuate in size and/or brightness are known as variable stars. Saiph brightness varies during 0.037 days from 2.021 (dimmest) to 1.986.

As of right now, there are no exoplanets in orbit around this star. They may or may not exist, but because of their size in relation to the star, they will be more difficult to see than other smaller stars, which explains why none have been seen near a supergiant.

Chapter Three

Orion's Belt: Alnitak, Alnilam, and Mintaka

Three large, very hot blue stars—Alnitak, Alnilam, and Mintaka—make up Orion's Belt. The stars are simple to recognise since they are uniformly spaced and roughly form a straight line. Three stars total, two of which are supergiants.

The stars are around the same age and generated in the same chemical cloud. The leftmost and rightmost stars in Orion's Belt, Alnitak and Mintaka, are around 1,200 light-years away from the Sun, but the centre star, Alnilam, is far further away. Approximately 2,000 light-years separate us from it. This indicates that, despite their apparent proximity to Alnilam in the sky, Alnitak and Mintaka are really closer to one another.

Alnitak

The leftmost star in Orion's Belt is Alnitak, Zeta Orionis (ν Ori). Within a triple star system, around 1,260 light-years distant, it is the main star.

The hot, bright blue O- and B-type stars make up the three elements of the Zeta Orionis system. Their total apparent magnitude is 1.77. The separate components glow with magnitudes 2.08, 4.28, and 4.01.

Formally, Zeta Orionis Aa, the main component, was called Alnitak. The Arabic term an-niṭāq, which means "girdle," is where the name originates. Al

Nitak or Alnitah was another spelling of it in the past.

The hot blue supergiant Alnitak belongs to the spectral class O9.5Iab. Its radius is 20 times that of the Sun, and its mass is 33 times that of the Sun. It glows with 250,000 solar luminosities at an effective temperature of roughly 29,500 K. It is assumed that the star is barely 6.4 million years old.

Alnitak ranks as the thirty-first brightest star overall and as the fifth brightest star in Orion. It outshines Mintaka while being somewhat fainter than Alnilam, its companion in the Orion Belt.

The brightest O-type star in the sky is Alnitak. The largest, bluest, hottest, and short-lived star types are known as O-type stars. They eat up their hydrogen supply more quickly than stars the size of the Sun due to their large mass. Alnitak is nearing the end of its life cycle, although having a fraction of the Sun's age. When it ends, it will explode as a magnificent supernova.

Alnitak and a blue subgiant star of the stellar classification B1IV form a near binary system. Alnitak A is another name for the binary star.

Zeta Orionis Ab (Alnitak Ab), the secondary component, has similarly reached the end of its main sequence lifespan but is less advanced. The star wasn't found until 1998.

It is also a potential supernova with a mass 14 times that of the Sun. Its radius is 7.2 times that of the Sun, and its surface temperature is 29,000 K. It is 32,000 times more luminous than the Sun. At 7.2 million years old, Alnitak Ab is somewhat older than its more gigantic neighbour.

The period of Alnitak Ab's orbit around Alnitak Aa is 2,687.3 days. The actual distance between them is just 11 astronomical units, or 35.9 milliarcseconds (Earth-Sun distances).

Even with the greatest telescopes, the secondary component is not visibly resolved. Only spectroscopic and interferometric methods may be used to identify it.

Zeta Orionis B, the third component, circles the main pair at a spacing of 2.728 arcseconds and has a period of 1,508.6 years. It is a B0III spectral class blue massive star.

Alnilam

The central star in Orion's Belt is Alnilam, also known as Epsilon Orionis (ε Ori). It is a lone star that is situated around 2,000 light-years distant. Star classification B0 Ia designates it as a bright blue supergiant.

Since Alnilam is the most massive and intrinsically luminous of the three Belt stars, it seems to be the brightest while being the farthest away.

Alnilam is the 29th brightest star in the sky and the fourth brightest star in Orion, with a magnitude of 1.69. Only Deneb is farther away from us among the top 30 brightest stars. Cygnus is 2,615 light-years distant from us.

Alnilam and Deneb (Alpha Cygni) belong to the same class of variable stars, the

Alpha Cygni variables. These are non-radial pulsating supergiant stars of the A and B types. Their surfaces are expanding in some areas while shrinking in others. The pulsations cause the brightness to change by around 0.1 magnitudes. Alnilam's brightness has been reported to range between magnitude 1.64 to 1.74.

The star's spectrum exhibits variability, maybe as a result of the significant mass loss it is going through. Compared to the Sun, Alnilam is shedding mass at a rate around 20 million times faster. Strong stellar winds, which may reach speeds of up to 2,000 km/s, are the reason for the mass loss.

The Arabic term al-niẓām, which means "the string (of pearls)," is whence the name Alnilam originates. Historically, it was also spelt Alnitam and Alnihan. The word "sapphire," "nilam," may have some connection to the name.

Alnilam has a radius of 32.4 solar radii and a mass between 40 and 44 times that of the Sun. At 27,500 K, its effective temperature, it is 537,000 times brighter than the Sun. It is thought to be 5.7 million years old.

Alnilam is a youthful star, yet it is already nearing the conclusion of its existence. Over the following million years, it may develop into a red supergiant more brilliant than Betelgeuse bcforc exploding as a supernova.

Mintaka

The star in the rightmost position in Orion's Belt, or leftmost when seen from the southern hemisphere, is Mintaka, also known as Delta Orionis (δ Ori). It is the 73rd brightest star in the sky and the seventh brightest star in Orion with an apparent magnitude of 2.23. It is the only star in Orion's Belt that is not a supergiant and is also the faintest.

The Arabic word manṭaqa, which means "belt," is the source of the name Mintaka.

The main member of a star system that is around 1,200 light-years distant is Mintaka. It has spectral type O9.5II, a hot blue brilliant giant.

The star has a mass 24 times greater than the Sun, making it a potential supernova. It is 16.5 solar radii in radius and 190,000 times brighter than the Sun.

The main component (Delta Ori Aa1) is a member of a triple star system, together with a B-type subgiant star (Delta Ori Ab) and a hot blue B-type main sequence star (Delta Ori Aa2). The closest companion, Delta Orionis Aa2, has a mass of 8.4 solar masses and a radius 6.5 times that of the Sun. It has an effective temperature of around 25,600 K and a luminosity of 16,000 solar.

Delta Orionis Ab has a radius of 10.4 solar radii and a mass 22.5 times that of the Sun. At 28,400 K, its surface is 63,000 times more brilliant than that of the Sun. It circles the two stars with a period of at least 400 years, and it is 0.26 arcscconds apart from the primary pair.

www.ingramcontent.com/pod-product-compliance
Lightning Source LLC
Chambersburg PA
CBHW072330270726
48658CB00016B/2255